LIFE SCIENCE INVESTIGATIONS

Under a Microscope
Small Life

HELEN LEPP FRIESEN

PERFECTION LEARNING®

Editorial Director: Susan C. Thies
Editor: Lori A. Meyer
Design Director: Randy Messer
Book Design: Emily J. Greazel and Robin Elwick
Cover Design: Michael A. Aspengren

A special thanks to the following for his scientific review of the book:
Paul J. Pistek, Instructor of Biological Sciences,
North Iowa Area Community College, Mason City, IA

Image Credits:

©Bettmann/CORBIS: pp. 5, 6 (left), 20; ©Ted Horowitz/CORBIS: p. 7 (top); ©Roger Ressmeyer/CORBIS: p. 7 (bottom); ©CORBIS: p. 16; ©Lester V. Bergman/CORBIS: pp. 24 (inset), 25 (left); ©Robert Pickett/CORBIS: p. 28

iStock International: pp. 3, 6 (bottom right), 9, 11, 14 (bottom), 15, 17, 18, 19 (top), 21 (inset), 22, 23 (left), 27 (inset); Perfection Learning: pp. 10, 12 (bottom), 14 (top); Photos.com: cover, pp. 1, 4, 8, 12 (top), 13, 19 (bottom), 21 (background), 23 (right), 24, 25 (top); 26, 27 (background), 31, 32

Text © 2006 by Perfection Learning® Corporation.
All rights reserved. No part of this book may be reproduced, stored in a retrieval system, or transmitted in any form or by any means, electronic, mechanical, photocopying, recording, or otherwise, without prior permission of the publisher.
Printed in the United States of America.
For information, contact

Perfection Learning® Corporation
1000 North Second Avenue, P.O. Box 500
Logan, Iowa 51546-0500.
Phone: 1-800-831-4190
Fax: 1-800-543-2745
perfectionlearning.com

2 3 4 5 6 7 PP 13 12 11 10 09 08

PB ISBN-10: 0-7891-6636-4 ISBN-13: 978-0-7891-6636-4
RLB ISBN-10: 0-7569-4697-2 ISBN-13: 978-0-7569-4697-5

Contents

1 The Microscope 4

2 What Are Living Things Made Of? 8

3 Bacteria 13

4 Mold 17

5 What's Up with Air? 21

6 What's in a Pond? 24

7 Soil's Secret Life 27

Internet Connections and Related Reading for Under a Microscope: Small Life 29

Glossary 30

Index 32

1 THE MICROSCOPE

Humans have always had a natural curiosity about unknown and unseen things. What is on the other side of a mountain, the world, or the universe? What's inside a human being, a plant, or an animal?

People invented horse-drawn carts, ships, automobiles, airplanes, and rocket ships to discover the vast, visible world. They invented the magnifying glass and the microscope to observe the insides of things and to see things invisible to the naked eye.

Invention of the Microscope

Around 1000 A.D. an unknown inventor made a reading stone. The inventor used the glass sphere to magnify, or enlarge, words in a book.

In 1590 two Dutch eyeglass makers, Zaccharias Janssen and his son Hans, conducted experiments with a variety of lenses in a tube. The multiple lenses caused objects to appear much larger than normal. This inspired the invention of the compound microscope and the telescope.

In 1665 Robert Hooke, an English physicist, noticed **cells** when he looked at a piece of cork under a microscope lens. His book, *Micrographia*, was an accurate and detailed record of his observations using a microscope.

Scientist of Significance

Anton van Leeuwenhoek (1632–1723)

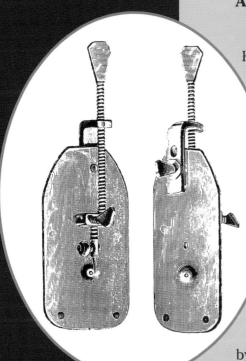

The first usable microscope made by Anton van Leeuwenhoek.

Anton van Leeuwenhoek was born in Holland. He was one of the first scientists to explore the invisible world. His hobby was making tiny glass lenses. He attached a lens to a hole in a metal plate and with this simple microscope examined his surroundings.

When Leeuwenhoek looked at pond water, he was amazed to see tiny creatures moving around in it. He saw one-celled organisms called *protozoa*. He examined blood and discovered that human red blood cells are round and those of fish and birds are oval. Leeuwenhoek also observed bacteria by scraping residue from his teeth and examining it.

Leeuwenhoek wrote about his discoveries for 50 years. His interest in the microscopic world led him to make a wide variety of microscopes.

Better Microscopes

Early scientists used compound microscopes, which bent light rays to form an enlarged image. These microscopes used two lenses and could magnify objects up to 1800 times.

In the 1930s scientists developed a new kind of microscope called the **electron microscope**. It could magnify objects over 100,000 times by shooting electrons in a beam instead of light at the object. This made it possible to view viruses.

Microscopes Today

There are several different types of microscopes used today. Each microscope has various functions and capabilities.

The dissecting microscope is similar to a fancy magnifying glass. It allows you to look at something in more detail than can be seen with the naked eye.

The compound light microscope has two lenses and uses light. This microscope has higher magnification and a higher resolution, or sharpness, than the dissecting scopes.

Scientists use an electron microscope in 1941.

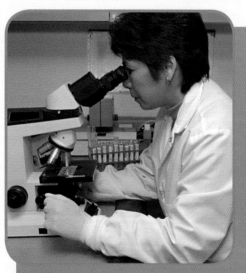
Compound light microscope

Today electron microscopes are often used because their resolution and magnification are much better than those of a light microsope. There are two kinds of electron microscopes—scanning electron microscopes (SEM) and transmission electron microscopes (TEM). The images seen in a scanning electron microscope are **three-dimensional**. The transmission electron microscope produces **two-dimensional** images.

A confocal microscope is a light microscope that uses lasers. The lasers scan an image that can be viewed on a computer screen. This microscope has the best resolution of any light microscope.

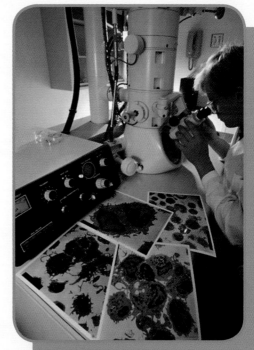
Transmission electron microscope

Today's microscopes are highly advanced. The magnification power of new microscopes can easily reach 200,000 times the size of the object, which is 10,000 times more than the earliest microscopes. Microscopes opened the door to the undiscovered, invisible world.

Confocal microscope

2 WHAT ARE LIVING THINGS MADE OF?

Your eyes are remarkable tools, but they do have limits. Try as hard as you can, but you still cannot see past your skin to the cells inside your body. When an object is less than 0.008 inches across, you cannot see it at all. Before microscopes were invented people had no idea that a whole invisible world existed around them and inside them.

Cells

A cell is the smallest part of any living thing that can carry out all the activities of life. Most cells are so tiny that you can't see them without the use of a microscope.

Some living things are made of millions of cells and others are made of just one single cell. Certain cells have a **nucleus** and other highly specialized compartments, while other cells have a very simple plan with no nucleus or compartments. Your body is made of millions of cells that work together to allow it to function smoothly.

Inquire and Investigate:
What Are the Building Blocks of a Plant?

Question: What are plants made of?
Answer the question: I think plants are made of _____.
Form a hypothesis: Plants are made of _____.

Test the hypothesis:
Materials:
- microscope
- knife
- tweezers
- iodine or blue ink
- onion

Note: An adult should use the knife to do the cutting. Iodine is a poison that stains skin and clothing.

Procedure:
Cut a fourth of a raw onion. Separate the sections of the onion. With the knife cut small squares into the onionskin and flesh underneath. With the tweezers, pick up one of the small squares. Put the piece of onion on a slide. Put a drop of iodine on it. Observe the slide. Draw what you see.

Observations: The onion cells are arranged like a brick wall. You can see the parts of each individual cell.

Conclusions: Plants are made of cells. Each cell has a cell wall, nucleus, cytoplasm, and cell membrane.

If you compare the cells of your body, you will find many different types of cells that look and act differently. Together, these cells make you who you are, a biological marvel. Even though cells may vary, there are qualities that all cells have in common.

Parts of a Cell

All cells, except for **bacteria**, have three main parts—the cell membrane, **cytoplasm**, and the nucleus. The cell membrane forms a "wrapper" around the cell and controls what materials can move in and out of the cell. The membrane also holds the cell together. The cytoplasm is the fluidlike mixture between the cell membrane and the nucleus where much of the activity of life occurs. It also contains the **organelles**, or parts that do special jobs.

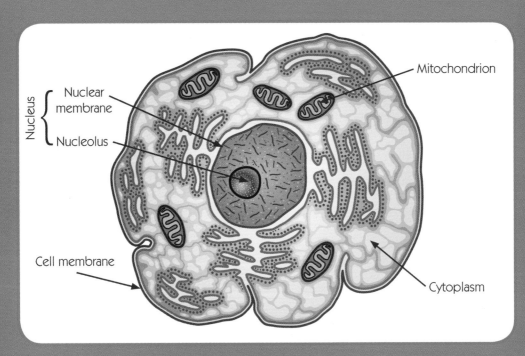

Animal cell

The nucleus is a large compartment found near the center of the cell. It serves as the "brains" of the cell and controls all of the cell's activities. The nucleus is where the **chromosomes** are located. Chromosomes contain genes, or instructions, that control how and when a cell divides, grows, and lives. The nucleus of a normal human cell contains 46 chromosomes.

DNA

A chromosome is made up of a **DNA** (deoxyribonucleic acid) molecule. DNA passes on characteristics, such as height or eye color, from an organism to its offspring.

Before a cell divides, its DNA makes a copy of itself so that both new cells will have the same set of instructions. DNA in the form of chromosomes can easily be seen under a microscope during cell division.

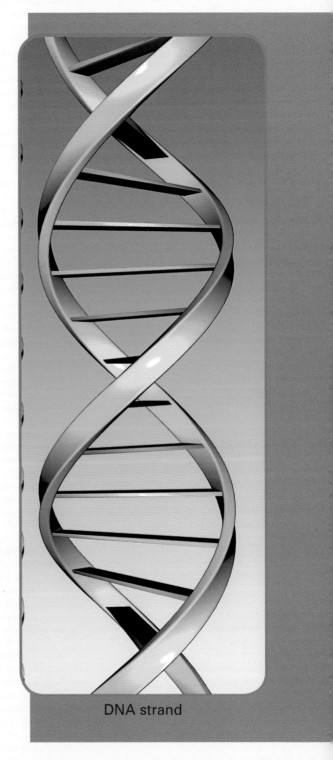

DNA strand

Decode the Message

DNA encodes messages in the body. Can you decode this secret message?

Message: # % ? ? ((! ~ %   ((" ! ? ?.

Key:

A !	B @	C #	D $	E %	F ^	G &
H *	I +	J <	K >	L ?	M "	N :
O =	P :	Q -	R ~	S ((	T)	U **
V ^^	W <<	X >>	Y ++	Z ##		

Answer: Cells are small.

Extra Parts

Plant cells have a cell wall, chloroplasts, and a large central vacuole. Animal cells don't have these parts.

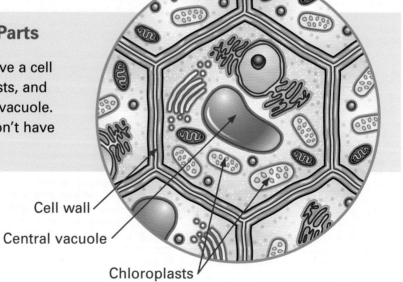

Cell wall
Central vacuole
Chloroplasts

Bacteria

3

What are bacteria? Bacteria are the smallest and most plentiful living creatures on Earth. They surround you, but you can't see them without a microscope.

Bacteria have one simple cell and reproduce by dividing in two. If their surroundings are good, some can reproduce every 20 minutes.

Inside Bacteria

Compared to other cells, bacteria's interior structure is quite simple. Bacteria are simple cells that don't have a nucleus. The outer surfaces of bacteria are made up of a cell membrane, cell wall, and sometimes a slime layer. The cell membrane, also known as the plasma membrane, regulates what moves in and out of the cell. Hairlike extensions of the cell membrane, called *pili*, are used for attaching the cell to another cell or a surface.

The cell wall serves as a support structure to maintain shape and to keep the cell from bursting under stress. The outer slime layer, when present, helps bacteria stick to surfaces and also prevents it from being eaten by other cells.

Bacteria have a threadlike extension called a *flagellum*. This extension is used to move bacteria through its watery environment. Flagella work somewhat like the propeller of a boat.

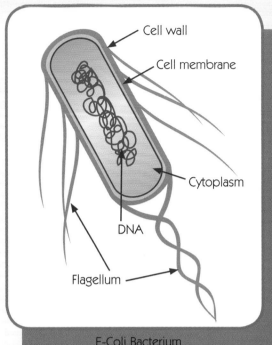

E-Coli Bacterium

A pink sinister bacteria with flagella

One or More?

Flagellum is singular or one and *flagella* is plural or more than one.

Bacteria Get Around

Bacteria live practically everywhere. They live on your skin, on your teeth, under your fingernails, on doorknobs, and on almost everything you touch. Bacteria live by absorbing food substances from their habitat. Some collect energy from sunshine.

Did you know that there are more bacteria on the body than the number of the body's own cells? Most bacteria live on the surface of the skin.

Bacteria also line the nose and mouth. You don't notice them because they are so small. The only time you do notice them is when they make you sick.

The Good and the Bad

Some bacteria are considered good and some are bad. Many bacteria are important for living things to survive. Good bacteria keep harmful bacteria under control. A good kind of bacteria is the bacteria in your digestive tract that produces certain vitamins. If bad bacteria get inside the body, they can cause an infection. Bad bacteria include the bacteria that cause diseases. These bacteria are called *germs*, or **pathogens**. Another bad kind of bacteria can get on food and cause food poisoning.

Scientists have classified about 10,000 different kinds of bacteria. They are classified by their shape. Each bacteria species is chemically different.

Observe This: Inside Your Mouth?

With a Popsicle stick, very gently scrape some of the soft white material that sticks to your teeth. Put it on a slide and add some saliva. Look at it under a microscope. Do you see the bacteria? Are they moving around?

Scientist of Significance:
Louis Pasteur (1822–1895)

Anton van Leeuwenhoek first observed bacteria under his simple microscope in the mid-1600s, but it wasn't until 200 years later that scientists realized that bacteria could cause disease.

Louis Pasteur was an accomplished scientist of the 1800s. With the use of a microscope, Pasteur discovered the bacteria that were spoiling wine. He conducted experiments and discovered that wine could be heated to kill the bacteria, and then it didn't spoil. This heating process is called *pasteurization*. Pasteur saved the wine industry of France using this process.

This heating process is used today to kill the bacteria present in raw milk. When you buy milk at the store, you will see that the label says the milk is pasteurized. Pasteurization of milk protects us from diseases.

Mold

4

Molds are microscopic **fungi**. They live on dead animals or plants, but some attack living matter. Molds may be gray, black, green, yellow, orange, or various colors, and may have a velvety or wooly texture.

Mold Growth

Molds grow in warm, dark, moist environments. Molds grow hyphae, which are thin individual threads, to feed on their **host** plants. When fungi attack living plants, they can kill them.

The fungi reproduce by growing **spore** cases at the end of the long, slender hyphae. The skin that covers the spore case pops open, and the spores fly away to settle on new surfaces to reproduce.

Mold spores

Observe This:
Grow Your Own Mold

Put moist newspaper at the bottom of a glass jar. Lean a piece of bread (without preservatives) against the side of the jar. Sweep the kitchen floor or under furniture to collect some house dust. Shake a little bit of house dust onto the bread. Close the lid of the jar. Keep it at room temperature, but out of direct sunlight. Observe the bread every day.

During the next few days, a white fluffy substance that looks like cobwebs will cover the bread. Do you see tiny black spots? When you do, open the jar, collect some mold, and examine it under a microscope.

Do you see the many threads attached to the spore cases? The black dots are the spore cases filled with many tiny spores. On the other end of the thread are the tiny, rootlike hyphae. They gather food from the bread.

Observe This: Fruit Mold

Put a moist piece of newspaper in the bottom of another glass jar. Instead of bread, place a slice of apple, orange, or peach in the jar. Sprinkle it with house dust and close the jar tightly. Let the jar sit in indirect sunlight and observe what happens. After a week observe the mold on the fruit. Examine the spore cases under the microscope.

Useful Mold

It may seem that molds are just out to destroy good bread or healthy fruit, but some molds are actually helpful. The mold *Penicillium* is a useful mold. Penicillium is used to make a medicine called *penicillin*. This antibiotic protects the body from infections and disease.

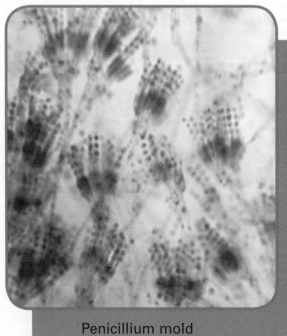

Penicillium mold

Technology Link:
The Discovery of Antibiotics

Antibiotics are molds that destroy or stop the growth of bacteria. Antibiotics are now important medicines.

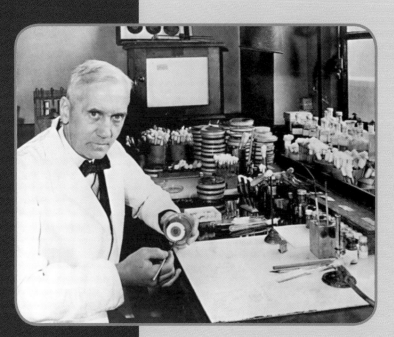

Alexander Fleming, a scientist who studied **microbes**, discovered penicillin by accident. In 1928 Fleming was conducting experiments in his laboratory at St. Mary's Hospital in London. He put a slate of disease-causing bacteria near an open window. When Fleming later observed the bacteria, he saw that a mold had blown in the window and spoiled the sample bacteria.

Fleming was surprised to see that the mold could dissolve the sample bacteria. Instead of throwing the spoiled specimen away, he researched this new aspect of his experiment. Other scientists joined the research of mold-killing bacteria and discovered penicillin. Penicillin became a lifesaving antibiotic that is still used all around the world today.

In 1945 Fleming, along with fellow researchers Howard Florey and Ernest Chain, won the Nobel Prize in Medicine for discovering penicillin.

What's Up with Air?

5

When you take a deep breath, you breathe in a lot more than just oxygen. If you see the Sun shining in your window in the late afternoon, you may see it reflecting off tiny particles floating around in the air. That is just a sample of what you breathe in.

What you breathe in depends on where you live. Living in the middle of a busy city puts a lot of smoke and dust in the air. You don't breathe that if you live out in the country where there is little traffic. If you looked at the air around you with a microscope, you'd see a whole city of objects with a variety of different shapes.

Allergies

Do you know anyone who has hay fever or **allergies**? People with hay fever or allergies often sneeze and have a runny nose and watery eyes. That's all due to what they breathe in.

The air that you breathe contains oxygen along with a mixture of other substances. Some of those substances are tiny grains of pollen, mold, and dust that some people are allergic to.

Pollen

Plants produce a generous amount of pollen. Wind shakes pollen from the stamen of a flower and carries it to the female part of the plant. That is how the plant reproduces. Unused pollen floats around in the air, and you breathe it.

It may be difficult to catch the pollen once it is floating in the air, but you can scrape it off the stamen of a flower. Pollen grains are the powdery yellow granules. Take pollen samples from different kinds of flowers and look at them under a microscope. How do they differ? Each plant has its own uniquely shaped pollen.

Fungal Spores

Fungus lives on dead plants. Fungi release spores into the air. The spores usually grow at the end of a long stalk and can easily

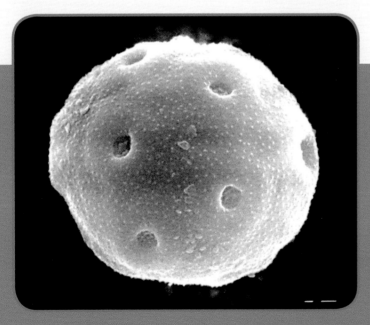

Electron microscope image of a pollen granule

be swept along by the wind. They have a simple structure made of only one cell covered by a resistant outer coat. This helps them survive in challenging weather conditions.

One or Two?

The plural form of *fungus* is *fungi*.
The plural form of *louse* is *lice*.

Wood louse

Book Lice

If you look carefully, you may see a book louse in one of your old books. Book lice are about 0.08 inches long, which means you can actually see them with the naked eye. They eat microscopic fungus that grows on the glue, binding, and paper of old books. Book lice lay eggs that stick to the pages and binding of old books.

Dust's Private Life

Dust has a life all its own. It is difficult to catch dust to look at all its different shapes and creatures. A good place to find dust is under your bed. If you can catch some dust, put it under a microscope and examine it.

Can you tell what is in the dust? Some of the things you might find are cloth fibers, hair, pet fur, insect scales, human and pet skin, dust mites and their droppings, and eggs of small animals. That's enough for anyone to say, "Pass the oxygen, please!"

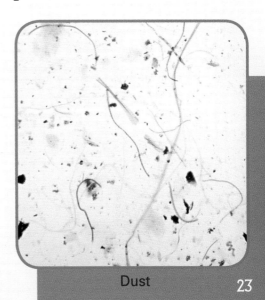
Dust

6 What's in a Pond?

If you look at a pond, you may see a frog hop here and a fish poke its head out over there. Otherwise the water may seem quite calm. Amazingly enough, a drop of pond water is actually brimming with life that you can't see. Many different types of invisible one-celled creatures called *protozoa* swarm around in pond water. These little creatures can move around, absorb and digest food, get rid of waste, and reproduce.

Protozoa

To identify the different types of protozoa, scientists have named them according to their shape and activity.

An *amoeba* is a gray protozoan that changes its shape as it moves around very slowly. An amoeba consists mainly of cytoplasm. It is barely visible to the naked eye.

Amoeba

A paramecium is a long, oval-shaped protozoan that scurries about quickly. Tiny hairlike structures called *cilia* cover the paramecium's surface and allow it to move. The paramecium feeds itself by sweeping tiny bits of food into its mouth-like structure and depositing it in a food vacuole. When the food vacuole fills up, it moves the food into its cytoplasm, where the food is digested.

One or More?

Protozoan is singular and *protozoa* is plural.
Paramecium is one and *paramecia* is more than one.

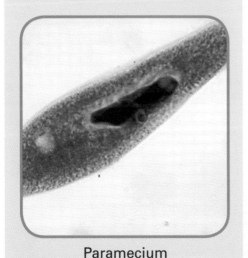

Paramecium

A large group of diatoms form a neat image.

Algae

The green slimy substance in pond water is called *algae*. These tiny plantlike cells grow connected as long strands or separately as single cells.

Diatoms are single-celled algae that have striking designs almost like a kaleidoscope. Their cell walls are made of silica, a material also found in sand. Diatoms make their own food.

Spirogyra is a type of algae that forms a strand. It has spiral-shaped chloroplasts. If you look at spirogyra under a microscope, you can see the different parts of the cell. It has a nucleus, cytoplasm, and cell wall.

Sample of Pond Water

To look at the fascinating world of pond water, collect a sample from the closest pond. Make sure your sample includes some of the green material growing in it and some of the dirt from the bottom of the pond.

Store your jar of pond water on a windowsill where there is light, but not direct sunlight. The heat might kill the specimen. Put a drop of water with the greenish material on a slide. What does the pond world look like? Do you see protozoa darting about? Do you see the long thin filaments of algae?

A pond contains many microscopic marvels.

Soil's Secret Life

If you let a handful of soil sift through your fingers, you may think soil is just millions of little grains of black dirt with a few earthworms slithering through it. The contents of soil are another well-kept secret. Some of these materials are visible, like the earthworm, but others are invisible. The use of a microscope lets you see into soil's secret life.

Life in Soil

Bacteria and fungi are found in soil. They survive by breaking down dead plants, insects, and animals. Some, like predatory fungi, feed on living beings. These fungi grow tiny nooses to trap little worms or other small prey in the soil.

Pseudoscorpions live in soil and feed off tiny insects in or around soil. They grab their prey with poisonous claws. Since they are tiny and can't move very far very fast, pseudoscorpions are known to hitchhike on the legs or backs of insects. These creatures are harmless to humans and only prey upon creatures of similar size. They can be found living in leaf remains on the ground. Pseudoscorpions are about 0.1 inches long.

Springtails live in the soil in plentiful numbers. They have to watch out for the pseudoscorpion, their predator. A unique characteristic of the springtail is its spring-loaded tail, which it keeps tucked underneath its body. When in danger the springtail flings itself to a safe place using its hidden tail.

A New Look

Since the invention of microscopes, scientists have opened our eyes to the once invisible world around us. We can take a peek into the makeup of cells and look beyond the surface of bread mold and pond water. Now anyone, with the aid of a microscope, can catch a glimpse of dust's private life and soil's secret life. It is an exciting world right at your fingertips, just waiting to be uncovered and discovered.

Springtail

Internet Connections and Related Reading for
Under a Microscope: Small Life

http://www.funsci.com/texts/index_en.htm A fun site with information about different kinds of microscopes.

http://inventors.about.com/od/mstartinventions/a/microscopes.htm Take a look at how microscopes evolved into the high-powered instruments of today.

http://www.howe.k12.ok.us/~jimaskew/mscope.htm This site tells all about light microscopes and includes some good illustrations.

http://www.eurekascience.com/ICanDoThat/bacteria_cells.htm Learn more about bacteria, plant and animal cells, and DNA.

http://www.hhmi.org/coolscience/airjunk/index.html An interesting look at what is in the air that you breathe, including an experiment to try.

Dirt: The Scoop on Soil by Natalie M. Rosinsky. Discusses the nature, uses, and importance of soil and the many forms of life that it supports. Picture Window Books, 2003. ISBN 1404800123. [IL K–4] (3429606 HB)

Greg's Microscope by Millicent E. Selsam. Tells the story of Greg and his microscope and the things he finds with it. HarperCollins, 1990. ISBN 006444144x. [RL 2.5 IL K–3] (4111001 PB)

In the Small, Small Pond by Denise Fleming. Explores the animals and activities in and around a pond. Henry Holt, 1998. ISBN 0805059830 (PB) 0805022643 (CC). [RL 3.1 IL 2–5] (6925801 PB 6925802 CC)

What's In the Pond? by Anne Hunter. This book tells all about life in a pond. Houghton Mifflin, 1999. ISBN 0395912245. [RL 4.6 IL K–4] (3349606 HB)

RL = Reading Level
IL = Interest Level

Perfection Learning's catalog numbers are included for your ordering convenience. PB indicates paperback. HB indicates hardback. CC indicates Cover Craft.

Glossary

allergies (AL luhr jeez) sensitivity to a substance that is harmless to most people

bacteria (bak TEAR ee uh) single-celled organisms without a nucleus

cell (sel) smallest unit of life

chromosome (KROH muh zohm) rodlike body of a cell nucleus that contains genes and divides when the cell divides

cytoplasm (SEYE tuh plaz uhm) fluidlike mixture of the cell that lies between the cell membrane and nucleus

DNA chemical material that makes up a chromosome and carries genetic information

electron (i LEK trahn) very small particle that has a negative charge of electricity and forms the part of an atom outside the nucleus

fungus (FUHN guhs) form of life that mostly feeds on nonliving (dead) remains of others; decomposer

host (hohst) living animal or plant that a parasite lives on

microbe (MEYE krohb) microorganism, mostly bacteria

nucleus (NOO klee uhs) compartment of nonbacterial cell that controls a cell's activities

organelle (or guhn NEL) part of a cell that carries out a specific job

pasteurization (pas chuhr uh ZAY shuhn) heating process that destroys bacteria

pathogen (PATH uh juhn) disease-causing agent, like some bacteria

protozoan (proh tuh ZOH uhn) microscopic, animal-like organism that is single-celled

spore (spohr) inactive, single-celled reproductive body of various plants, fungi, and bacteria that can produce a new individual

three-dimensional
(three duh MEN shuh nuhl)
figure that has height, width, and depth

two-dimensional
(too duh MEN shuh nuhl)
figure that has length and width, but no depth

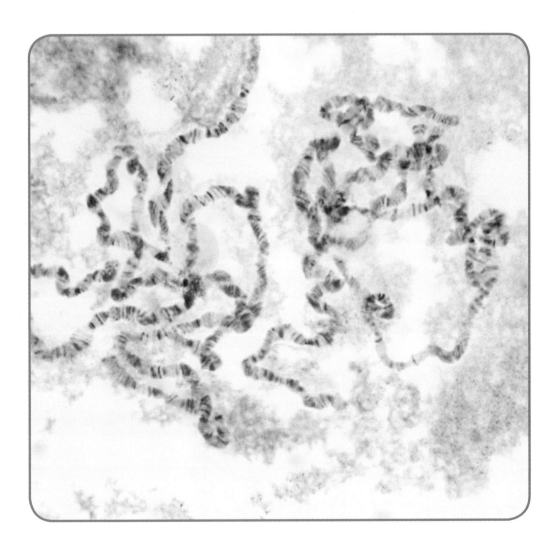

Index

allergies, 21
amoeba, 24
antibiotics, 19
book lice, 23
cells, 5, 8–12
  animal, 10–11
  plant, 12
Chain, Ernest, 20
cytoplasm, 10
diatom, 25
DNA, 11–12
flagellum, 14
Fleming, Alexander, 20
Florey, Howard, 20
fungus, 17, 22
Hooke, Robert, 5
Janssen, Zaccharias, 4
Leeuwenhoek, Anton van, 5, 16
Micrographia, 5

microscope, 4–7
  compound, 5–6
  confocal, 7
  dissecting, 6
  electron, 6–7
    scanning electron, 7
    transmission electron, 7
mold, 17–19
Nobel Prize, 20
nucleus, 8
paramecium, 25
Pasteur, Louis, 16
penicillin, 19–20
pollen, 22
protozoan, 5, 24–26
pseudoscorpion, 28
spirogyra, 26
spore, 17
springtails, 28

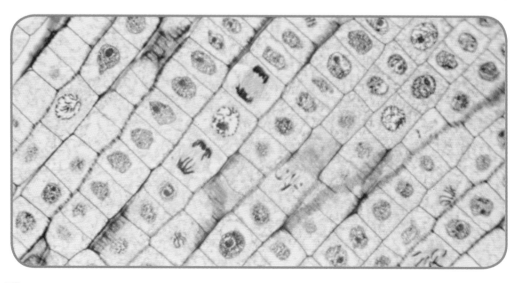